Mathematical Olyn

for

Elementary School 1

My First Book of Mathematical Olympiads – *First Grade*

(Workbook)

My First Book of Mathematical Olympiads

Mathematical Olympiads *for* Elementary School

1

First Grade

(Workbook)

Michael Angel C. G., Editor

Preface

The *Mathematical Olympiads for Elementary School* are open mathematical Olympiads for students from 1st to 4th grade of elementary school, and they have been held every year in the city of Moscow since 1996, their first editions taking place in the facilities of the Moscow State University - Maly Mekhmat. Although initially these Olympiads were conceived for students of a study circle of elementary school, then it was extended to students in general since 2005. Being the Technological University of Russia – MIREA its main headquarters today. Likewise, these Olympiads consist of two rounds, a qualifying round and a final round, both consisting of a written exam. The problems included in this book correspond to the final round of these Olympiads for the 1st grade of elementary school.

In this workbook has been compiled all the Olympiads held during the years 2011-2020 and is especially aimed at schoolchildren between 6 and 7 years old, with the aim that the students interested either in preparing for a math competition or simply in practicing entertaining problems to improve their math skills, challenge themselves to solve these interesting problems (recommended even to elementary school children in upper grades with little or no experience in Math Olympiads and who require comprehensive preparation before a competition); or it could even be used for a self-evaluation in this competition, trying the student to solve the greatest number of problems in each exam in a maximum time of 1 hour. It can also be useful for teachers, parents, and math study circles. The book has been carefully crafted so that the student can work on the same book without the need for additional sheets, what will allow the student to have an orderly record of the problems already solved.

Each exam includes a set of 8 problems from different school math topics. To be able to face these problems successfully, no greater knowledge is required than that covered in the school curriculum; however, many of these problems require an ingenious approach to be tackled successfully. Students are encouraged to keep trying to solve each problem as a personal challenge, as many times as necessary; and to parents who continue to support their children in their disciplined preparation. Once an answer is obtained, it can be checked against the answers given at the end of the book.

Sincerely,

The editor

Contents

Problems

Olympiad 2011

(XV Olympiad for Elementary School)

Problem 1. The hedgehog has a 4-wheel bike, Krosh has a 3-wheel bike, and Nyusha has a 2-wheel bike. Krosh, Nyusha, and the hedgehog went for a walk, some on bikes, some on foot. Losyash counted the number of wheels: it turned out 7. Who went out for a walk on foot?

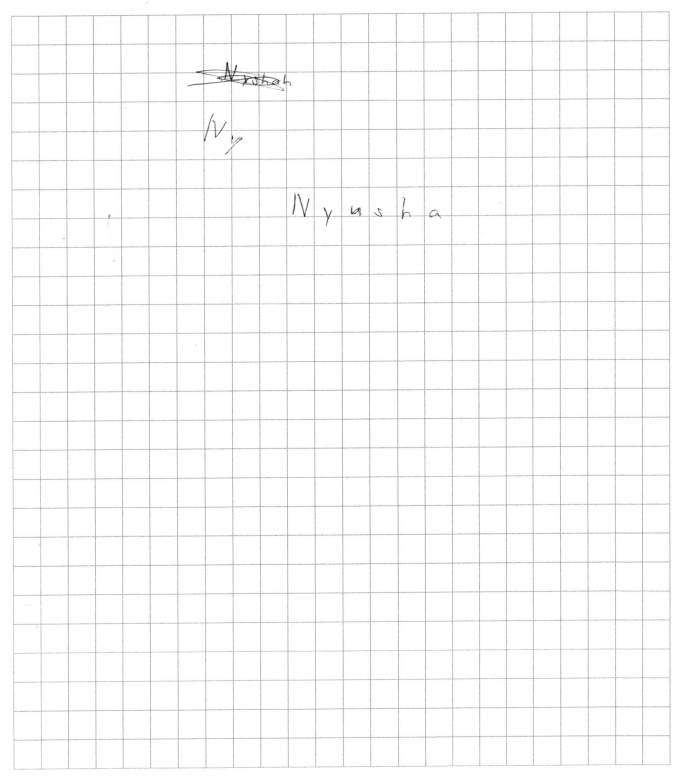

Problem 2. A portion of a large gift-wrapping paper has been cut out. Petya found 6 different pieces. Which of the pieces belongs to this paper?

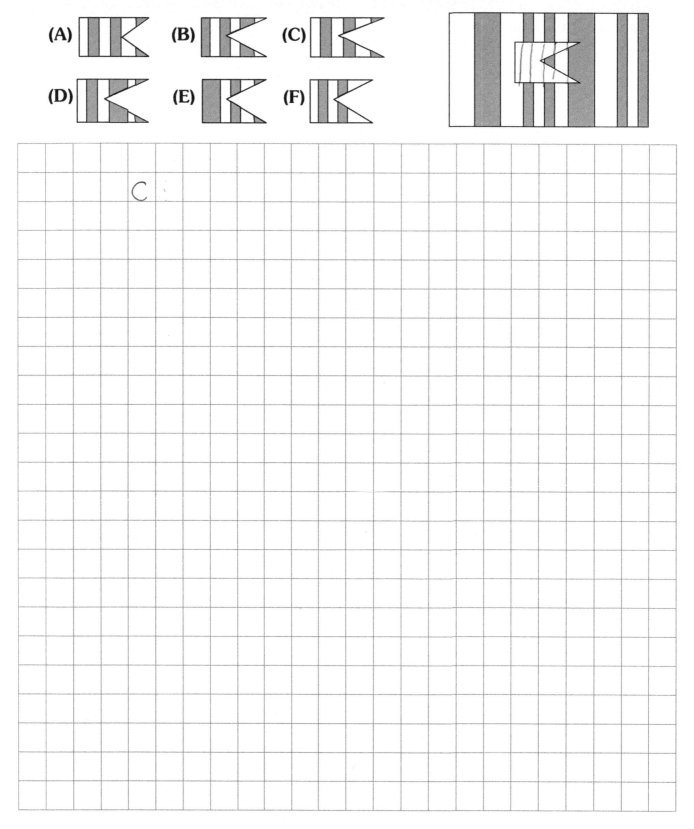

Problem 3. Gosha folded square paper in half and then in half again. Then, he pricked the resulting square in the middle with a fork. How many holes will Gosha see when he unfolds the square?

4 × 4 = 16

Problem 4. In a 4-cell square, Olya painted more than 3 cells with a yellow pencil, and then Ira painted more than 3 cells with a blue pencil. It turned out that all the cells in the square are painted. Also, if you paint the cell with yellow and blue pencils, you get green. How many green cells are in the square?

Problem 5. Cut the figure shown in the picture into three equal pieces.

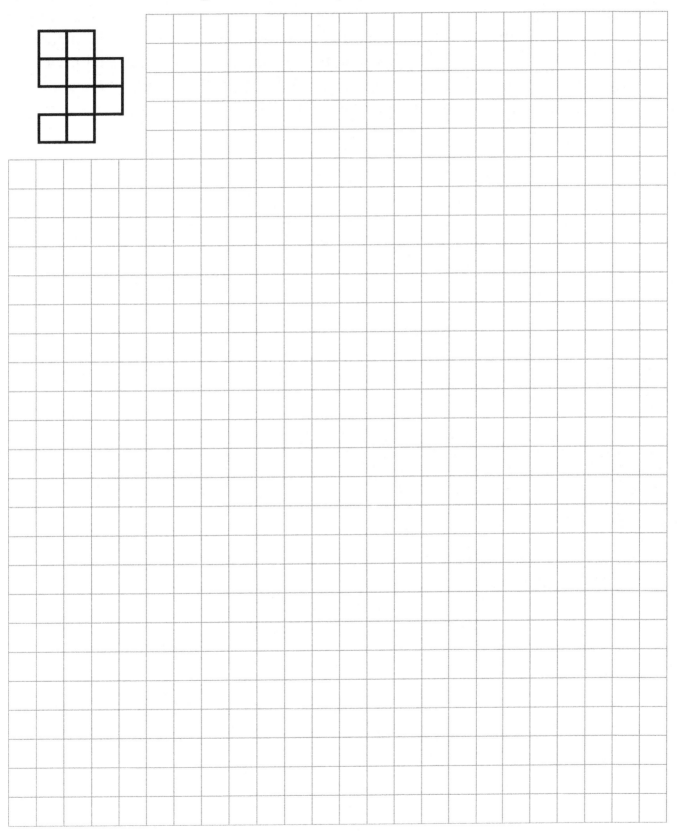

Problem 6. In the grove, in the thicket and on the borders of the forest there are mushrooms: milkcap mushrooms, poplar mushrooms and honey mushrooms. And in each place there is only one species of mushroom. Petya went to gather mushrooms in the thicket and grove, and brought milkcap mushrooms and poplar mushrooms. Vasya went to the grove and the borders of the forest and returned with poplar mushrooms and honey mushrooms. If Dima goes for the milkcap mushrooms. Where should he go?

Problem 7. Winnie the Pooh, Piglet, Donkey Eeyore and Christopher Robin play in a seesaw (*see* figure). It is known that Winnie the Pooh and Piglet together weigh more than Donkey Eeyore, and that Eeyore weighs more than Christopher Robin and Piglet together. Who weighs more if Christopher Robin and Winnie the Pooh sit on the seesaw?

Problem 8. "I love oranges", Sonya said. "No, I love oranges. You love apples", Vera said. "I don't like oranges", Andrei said. It is known that there is at least one child who likes each fruit and that no child likes both fruits at the same time. Which child loves apples and which child loves oranges, if they all lie?

Olympiad 2012

(XVI Olympiad for Elementary School)

Problem 1. My sister's name is Anna Pavlovna. My mother's name is Svetlana Dmitrievna and my grandfather's name is Ivan Petrovich. What is my father's name?

Problem 2. Cut the checkered figure shown below into two equal parts.

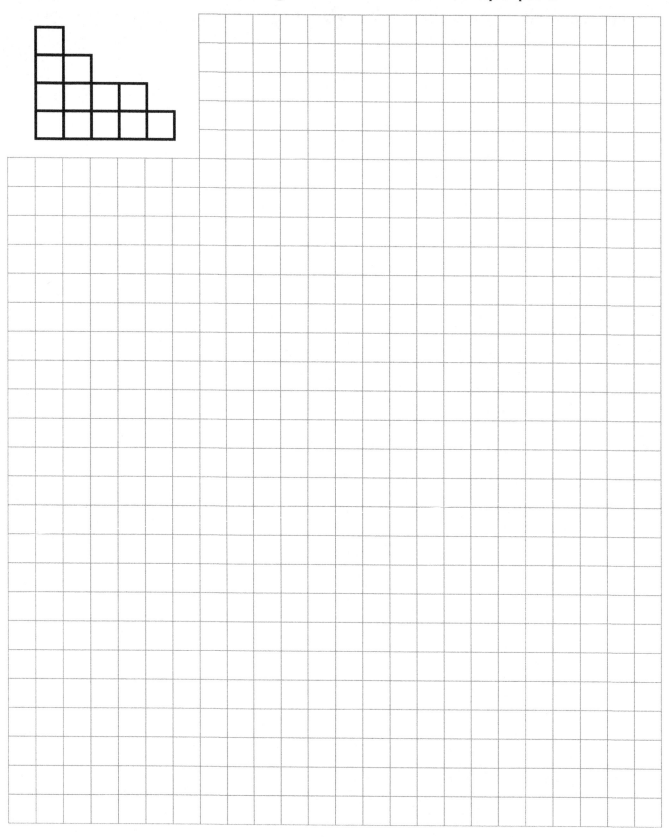

Problem 3. Andryusha photographed reflections of fruits in the mirror. And then he lost a picture. Which picture is Andryusha missing now?

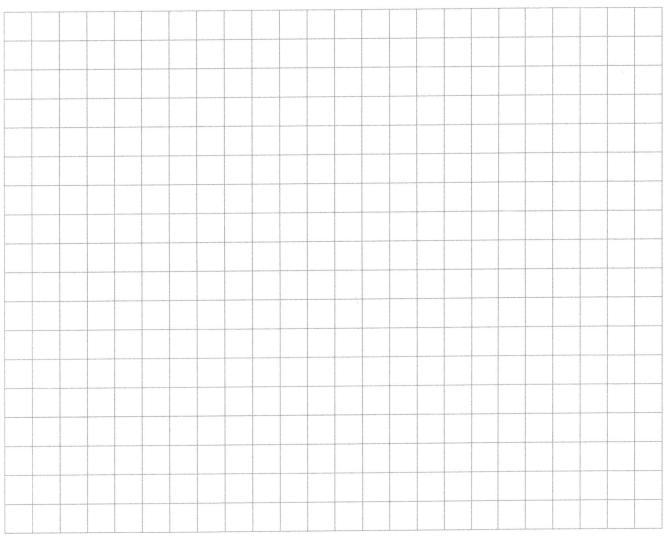

Problem 4. In the table below, arrange the circles, triangles, squares, and crosses so that in each column and each row, as well as each highlighted small square, there are all four figures.

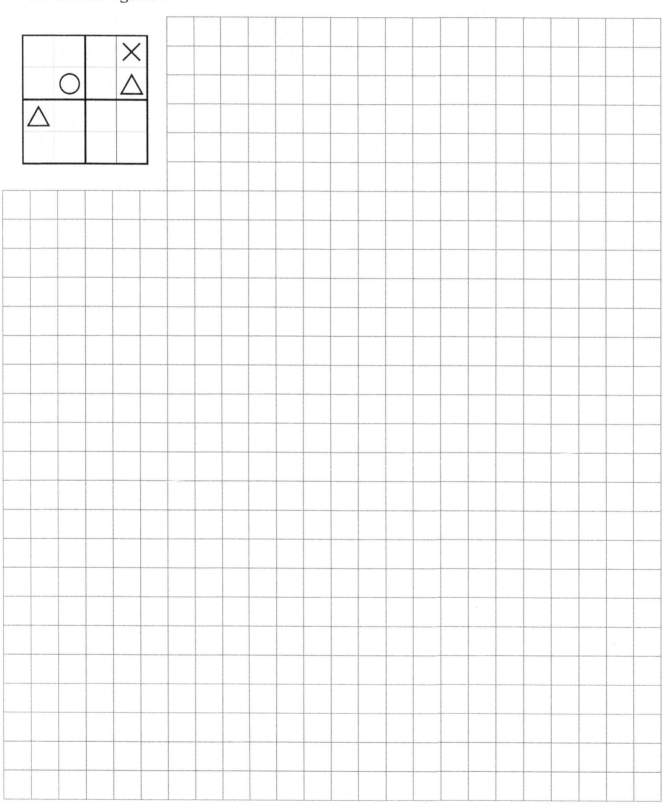

Problem 5. Three cars participating in a race came out in this order: yellow, red, blue. They reached the finish line in this order: "Honda", "Mercedes", "Audi". At the same time, no car finished the same way it started. What color is each car brand, if Audi is not yellow?

Problem 6. Paths have been built in the Enchanted Forest (see picture). It turned out that the path from Piglet to Eeyore Donkey is 7 *km*, the path from Winnie the Pooh to Rabbit is 4 *km* and from Piglet to Winnie the Pooh is 3 *km*. How many kilometers will someone have to travel down the path from Eeyore to Rabbit?

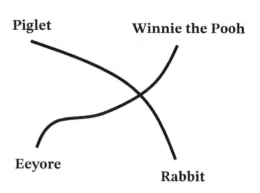

Problem 7. Petya goes up from the first floor to the fourth in 4 minutes. And Masha from the fourth floor to the seventh, in 3 minutes. Which of them will go the fastest from the first floor to the seventh and in how many minutes?

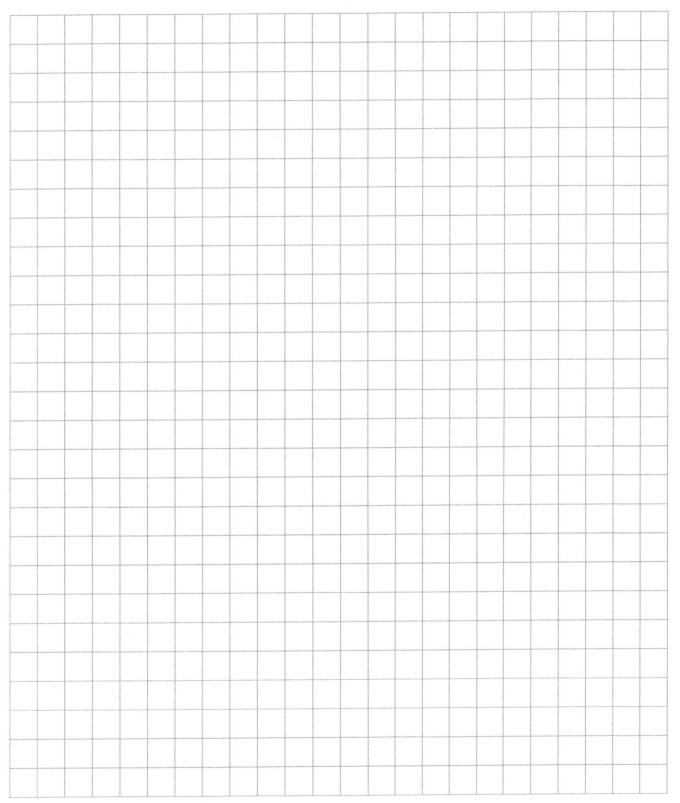

Problem 8. Twins Misha and Grisha lie at the same time only on Sunday. Other days, one lies and the other tells the truth. Misha said, "Today is Sunday". Grisha replied: "Sunday is tomorrow". What weekday is today?

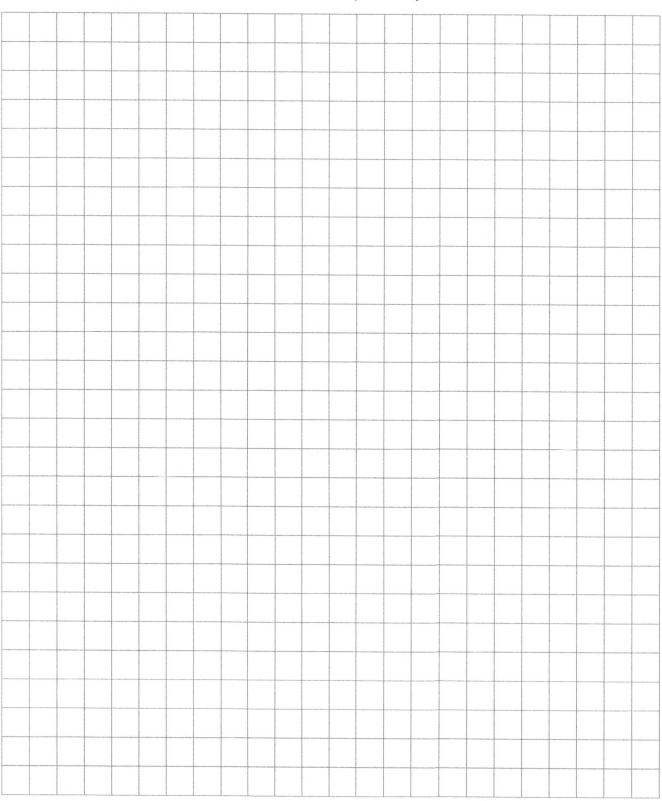

Olympiad 2013

(XVII Olympiad for Elementary School)

Problem 1. Mowgli, Bagheera the panther, Kaa the boa constrictor, Baloo the bear, and two wolf cubs met at the forest glade. How many legs are there in total?

Problem 2. The penguin is building an ice house. It remains to insert a part. Which?

Problem 3. Cut the pretzel in the picture into 6 pieces with two straight cuts.

Problem 4. Martians have three arms. The group from the Martian kindergarten lined up in pairs for a walk (see figure) According to the rules, each child must take each of her neighbors by the hand (left, right, forward or backward). How many hands free will the children in this group have after doing this?

Problem 5. The ancient Romans, instead of the usual numbers 1, 2, 3, ... wrote the numbers in a different way: instead of 1 they wrote I, instead of 2 - II, instead of 3 - III, instead 4 - IV, instead of 5 - V, instead of 6 - VI, instead of 7 - VII. Vasya wrote the Roman numeral "5" on the sheet of paper and Julia wrote the Roman numeral "4". They went with their sheets of paper to the mirror. Who has the highest number in the mirror: Vasya or Julia?

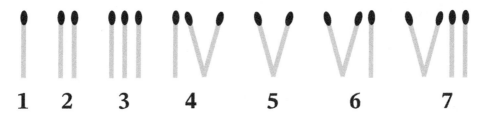

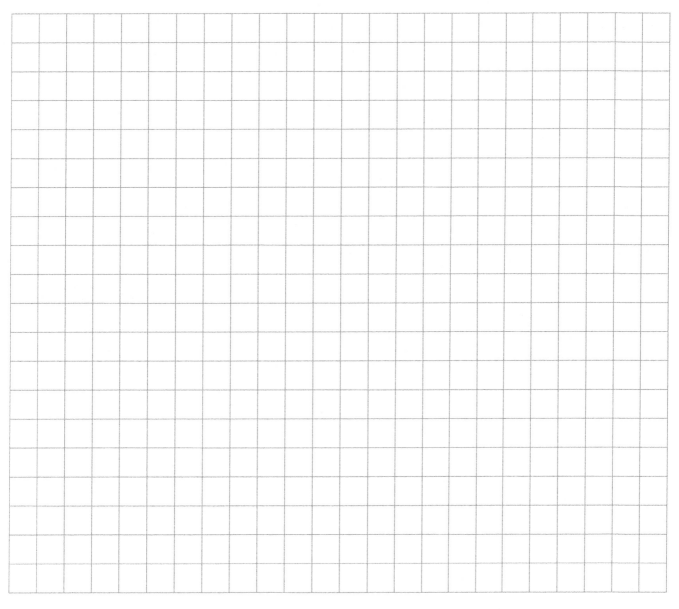

Problem 6. Cheburashka runs every morning in a park with five oak trees. Draw the Cheburashka route and indicate in what order he passes the oaks, if he runs exactly once on each path. The direction of Cheburashka's movement is shown by arrows.

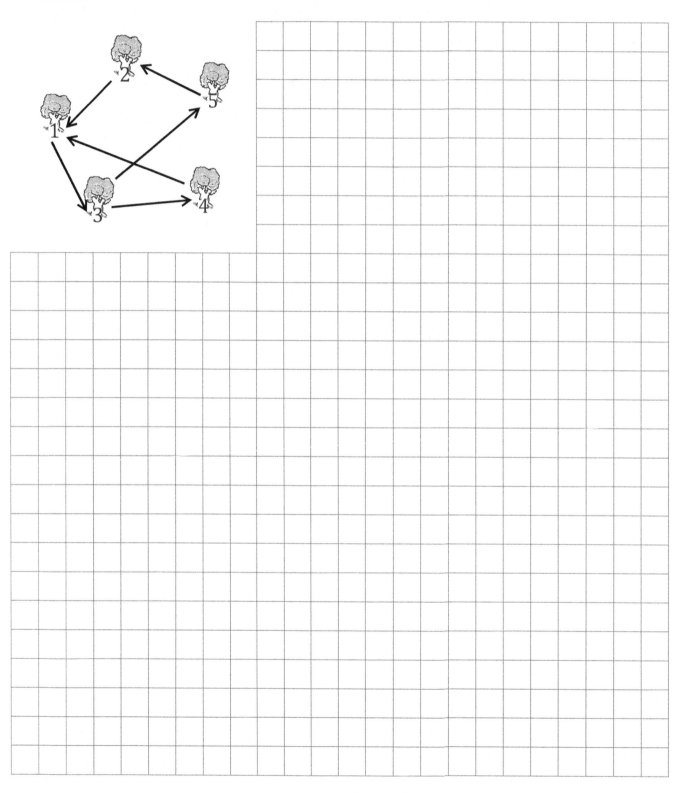

Problem 7. In a forest a group of mushrooms grow in a circle. In the forest glade, russulas always grow between two fly agarics. Vitalik counted without omitting anything, thus: "Two russulas, one fly agaric, seven russulas, one fly agaric". And Egor from another part of the circle: "Three russulas, a fly agaric, five russulas, a fly agaric". How many russulas are in the glade?

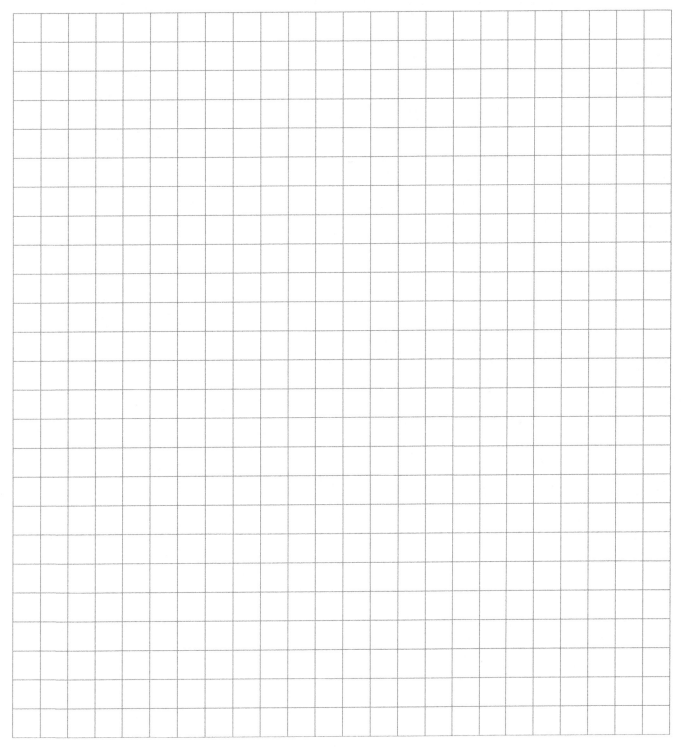

Problem 8. Krosh, Hedgehog, and Nyusha draw lots (three sticks of different lengths) for who would drive the car. Krosh said, "I have the shortest stick!" Nyusha said, "I'll drive!". Who drove the car if the one with the shortest stick did it and if both Krosh and Nyusha lied?

Olympiad 2014

(XVIII Olympiad for Elementary School)

Problem 1. Petya has apples and Sasha has pears. If Petya trades one of her apples for two of Sasha's pears, then they will have the same number of fruits. How many more pears than apples are there?

Problem 2. Five gears are meshed with each other and rotate. If the upper left gear rotates clockwise as shown in the figure. In which direction does the lower left gear rotate?

Problem 3. On television, two episodes of a movie are broadcast on different channels on the same day. The first episode three times - at 12:00, 17:00 and 18:00 on the channel "AGA", and the second episode also three times - at 11:00, 15:00 and 17:00 on the channel "OGO". Egor wants to see the first episode first, and then the second. When should you turn on the television if each episode is exactly one hour long?

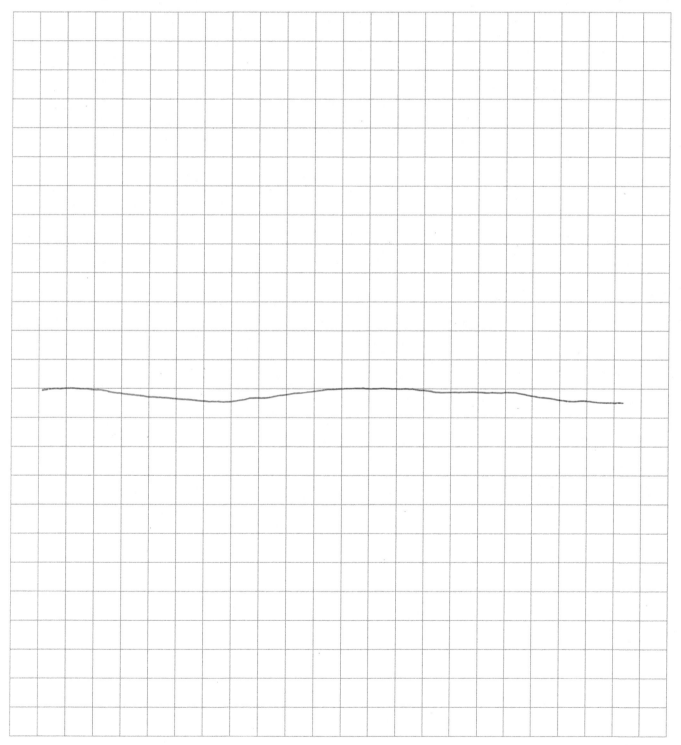

Problem 4. Gosha folded a square paper in half, then folded it in half again. Then he cut the resulting folded paper with two cuts, as in the figure. How many pieces of paper will Gosha find after unfolding it?

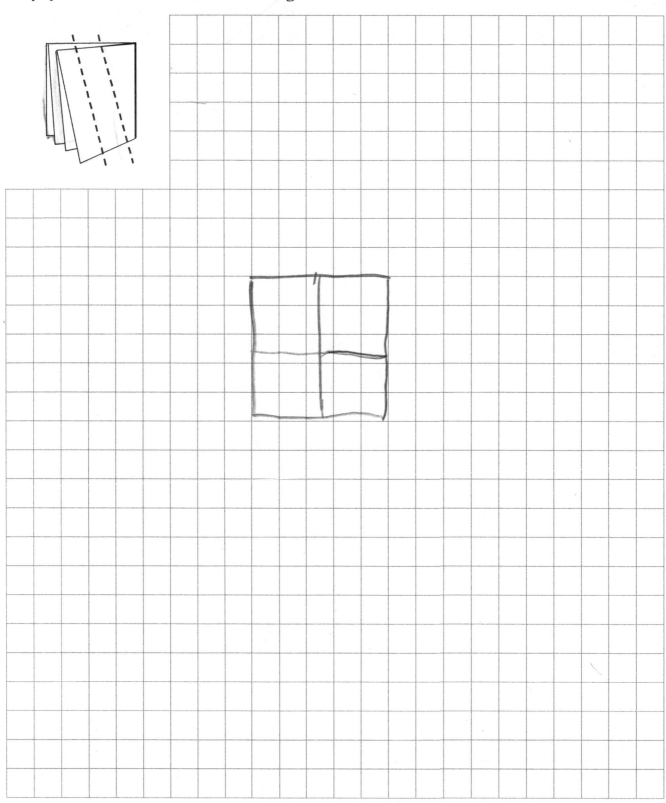

Problem 5. Masha, Dasha and Sasha go to school and are in the same class. On the night of September 1, they were asked the name of their teacher. "Maria Mikhailovna," Masha said. "Katerina Mikhailovna," Dasha said. "Alisa Stepanovna," Sasha said. It turned out that each of them correctly remembered only the first name or only the middle name. What is the teacher's name?

Alisa mikhailovna

Problem 6. Passengers in 3 cars of a train carry 7 kittens, the third car has the least number of kittens and the second car has twice as many kittens as in the first car. How many kittens are in each car?

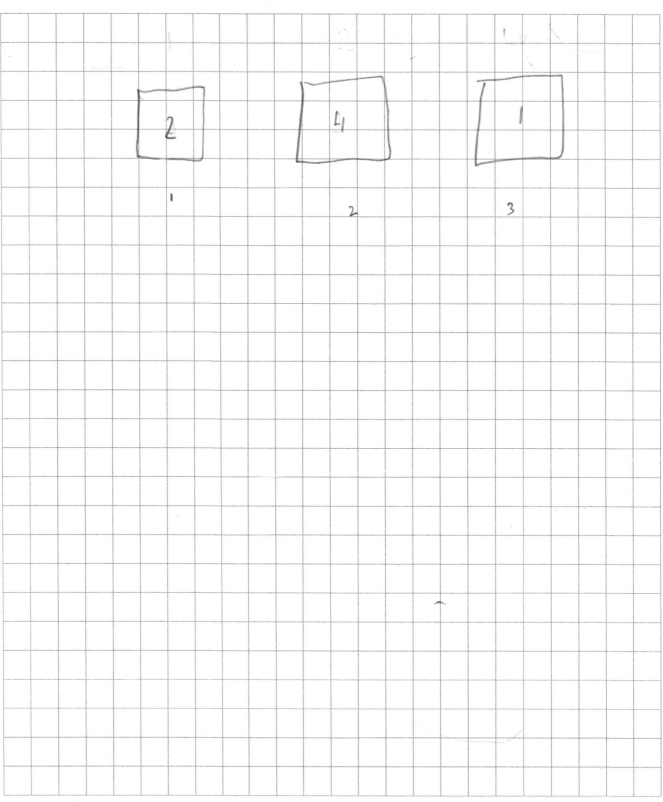

Problem 7. Losyash came up with building numbers with dominoes. For example,

is the number "21" and is 2350. Build the largest number

possible with dominoes y .

6 2 4 3

Problem 8. Mutta, Mokhovaya Beard and Polbotinka were eating ice cream. "Polbotinka ate more than everyone!" - Mutta said. "No, I ate less than Mokhovaya Beard", Polbotinka objected. "Polbotinka and I ate alike", Mokhovaya Beard said in a conciliatory tone. Who ate less if it is known that everyone lied?

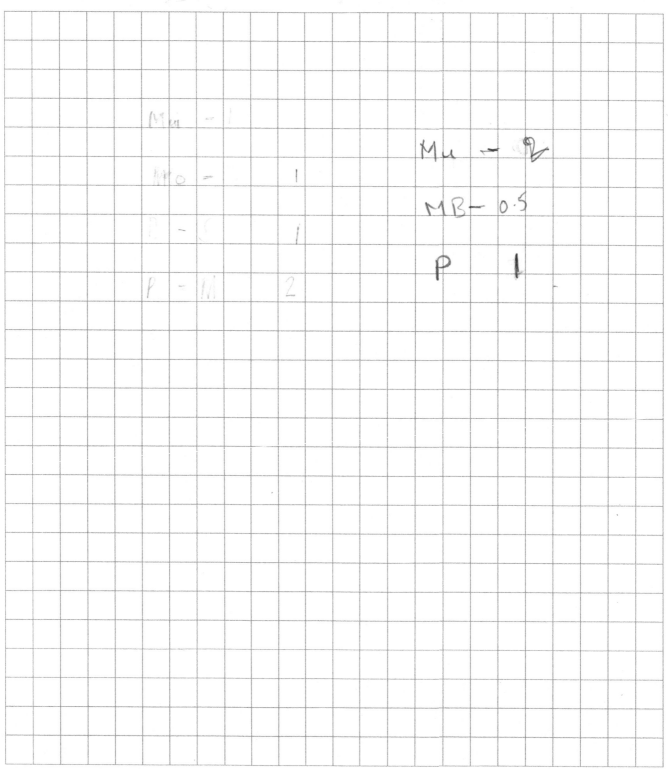

Olympiad 2015

(XIX Olympiad for Elementary School)

Problem 1. Now Tanya, Mana and Anya are 12 years old in total. How old will they be in 2 years?

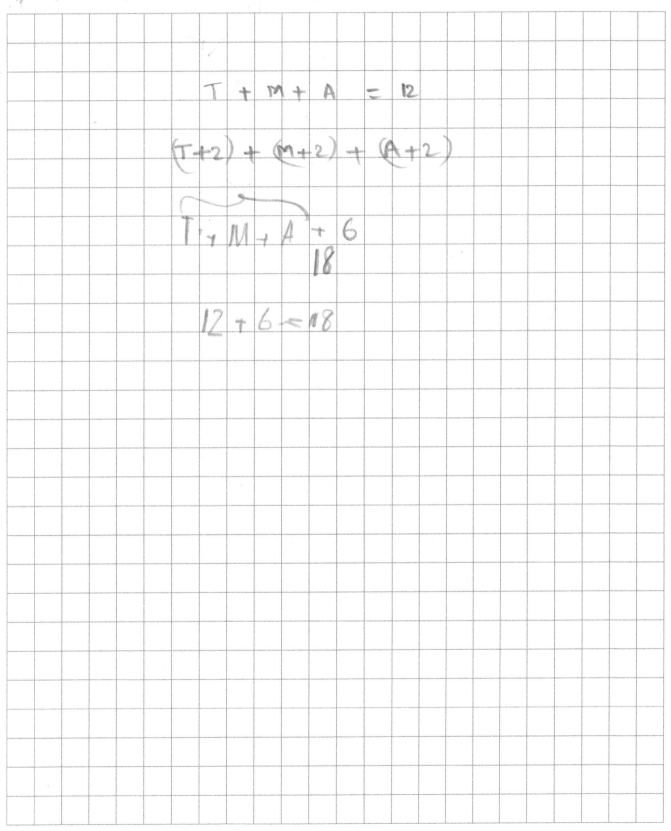

$$T + M + A = 12$$

$$(T+2) + (M+2) + (A+2)$$

$$T + M + A + 6$$
$$18$$

$$12 + 6 = 18$$

Problem 2. In the number 798, all the numbers are different. What number closest to this has the same property?

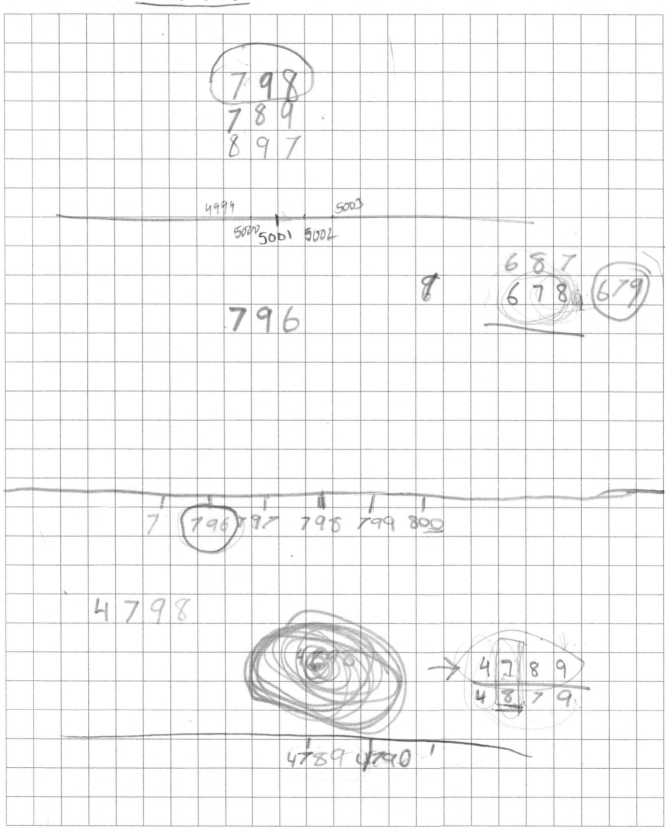

Problem 3. Petya brought his dachshund dog Dina to the dog show. If he photographed the dachshunds in the order the spots were assigned. Which picture is of Dina? (Petya was photographing against the sun, so only silhouettes were obtained.)

Dina

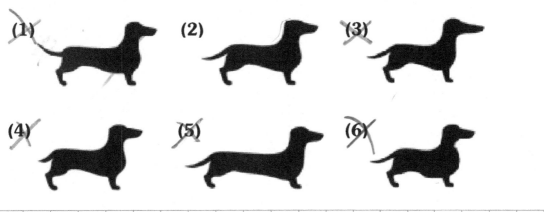

Problem 4. Draw 10 shapes in a row: circles and squares so that next to each circle there are only squares, and next to each square there is a circle and a square.

Problem 5. Yegor sketched a mountain with a layer of snow with the help of matches. rearrange 2 matches so that exactly 3 triangles are visible in the figure (there should be no additional matches).

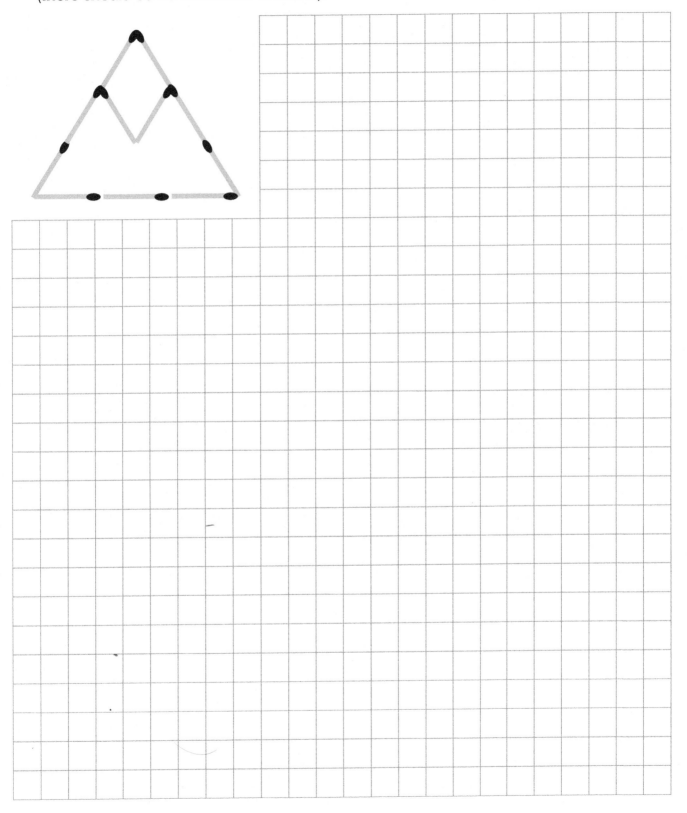

Problem 6. Vanya received a 9-cell chocolate bar as shown in the figure. If you want to eat two cells of the chocolate bar so that the rest does not fall apart. How many ways could she do it?

Problem 7. A beetle moves along a 1 meter long ribbon (as in the picture). It begins to move from the starting point 2 *cm* from one end of the ribbon and moves strictly in the middle of the ribbon, without going sideways or going back. How much distance will the beetle travel when it moves toward the arrival point 3 *cm* from the other end of the ribbon?

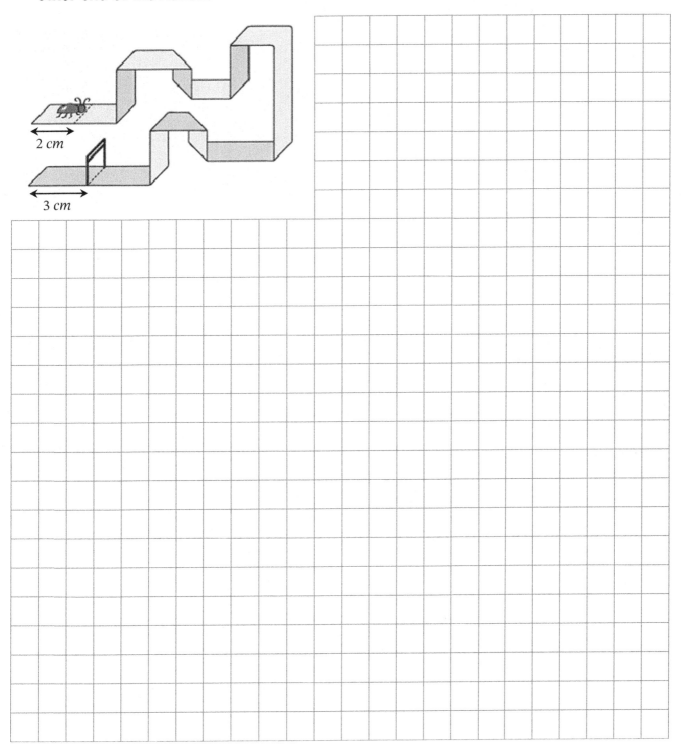

Problem 8. Vasya, Grisha and Dima participated in a car race with three cars: blue, red and yellow. Dima's car reached the finish line right after the yellow car and the red car, right after Vasya's car. Whose car and what color was the car that arrived first, if not Grisha's?

Olympiad 2016

(XX Olympiad for Elementary School)

Problem 1. A mouse made a big hole in a sweater. Baba-Manya wove various patches. Which is the most suitable for mending the sweater?

Problem 2. I have three friends: Anton, Borya and Kolya. Yesterday I played with Borya and Anton. One of them is 8 years old and the other 9 years old. And today I walked with Anton and Kolya. One of them is 10 years old and the other 8 years old. How old is each of my friends?

Problem 3. Instead of the asterisks, put some numbers to get the correct equality

$$2 * - * - * = 3$$

Problem 4. A beetle is in the cell of a board (as in the figure). If he walked through 2 cells and stopped at the third. Indicate which cell it could be in.

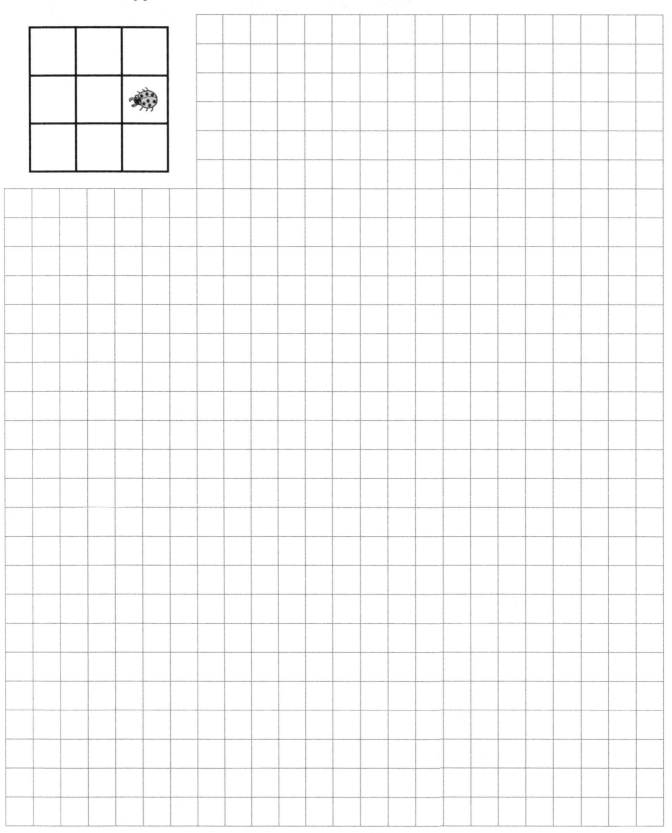

Problem 5. A watermelon was cut with three straight cuts as shown in the figure. How many pieces of peel were obtained after eating the watermelon?

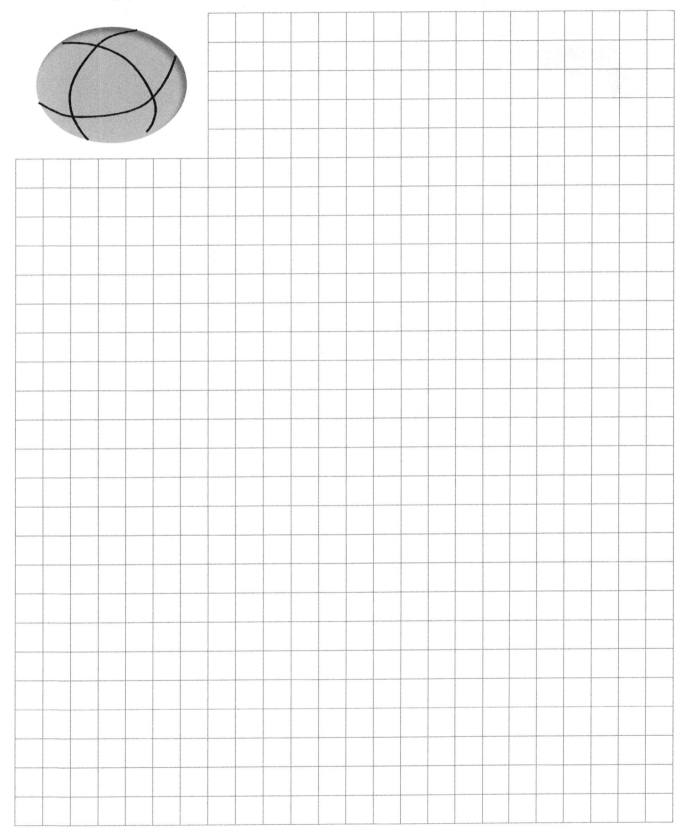

Problem 6. Vasya, Gosha, and Kasimir drew a white triangle, a gray circle, and a black square with oil paints (each child drew a shape). It is known that Vasya drew later than Kasimir, and Vasya's shape is not white. Who drew what shape?

Problem 7. Some children came up with the idea of adding numbers with matches:

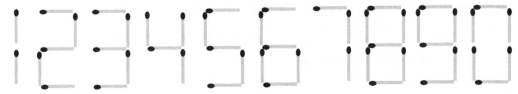

Anya posted the following wrong equality. Move 2 matches so that the equality is true:

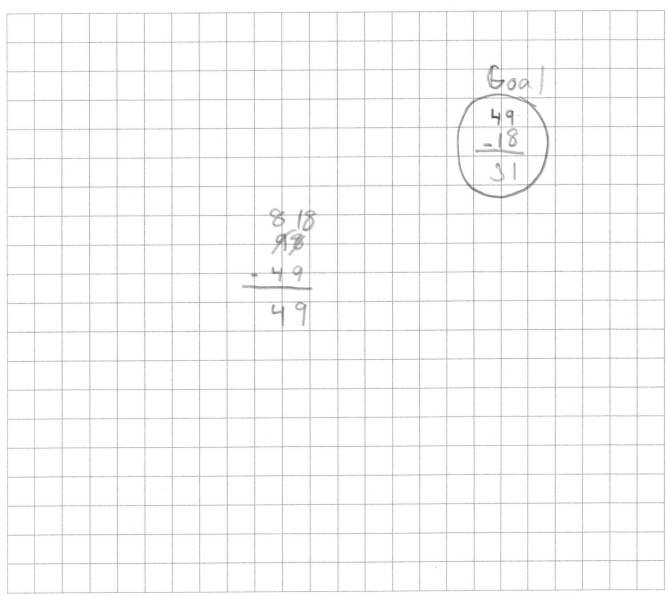

Problem 8. Three friends Tikhon, Yegor and Vitalik exchanged toys. Vitalik started playing with a fire truck. The owner of the dump truck liked the excavator. The owner of the fire truck took the dump truck. Determine which toy is whose, if it is known that the fire truck is not Tikhon's.

Olympiad 2017

(XXI Olympiad for Elementary School)

Problem 1. Petya drew a maple leaf like the one shown in the figure opposite. Which of the following drawings did Petya make?

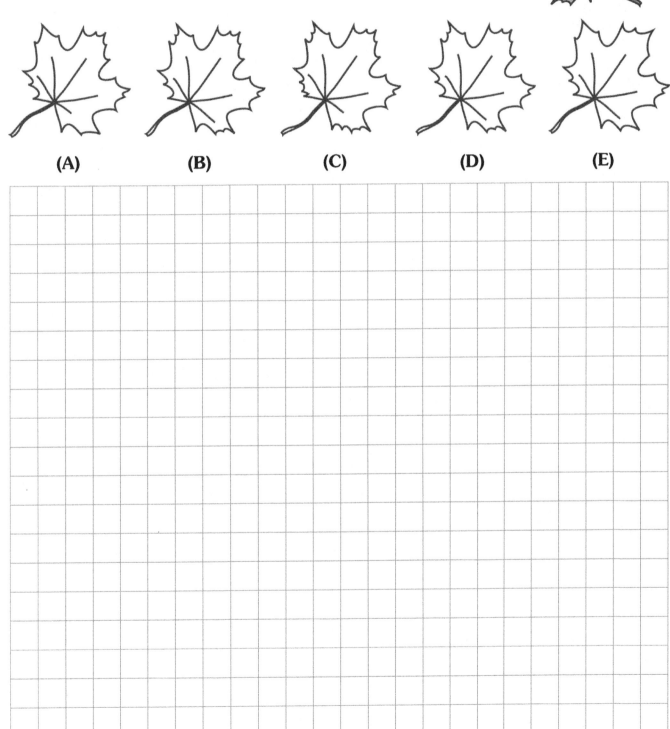

(A)   (B)   (C)   (D)   (E)

Problem 2. Vasya spends more time eating breakfast than Petya brushing his teeth and washing his face. And Murka spends the same time washing up as Vasya eating breakfast. Who will finish washing up faster, Murka or Petya?

Problem 3. Replace the letters with numbers from 1 to 7, so that the inequality is true. If different letters represent different numbers.

$$S < N < E > W < I < N > K > A$$

Problem 4. Masha, Katya, Tikhon, Yegor and Sveta line up for ice cream. It is known that Yegor is in front of Tikhon and Katya is behind Masha and in front of Sveta. What is the order of the children in line if there are no two girls together?

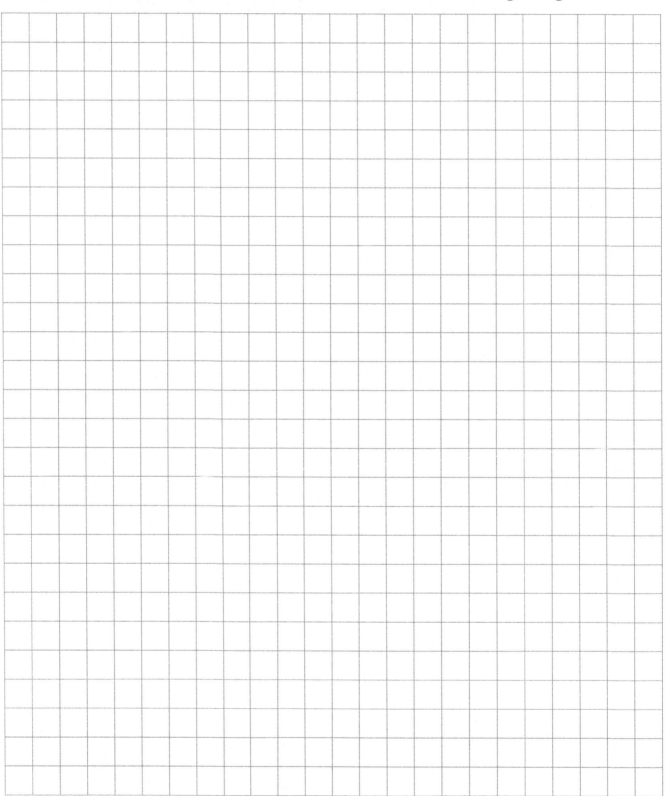

Problem 5. Krosh has two clocks. One of them is 2 hours ahead, and the other is 1 hour behind. What time is it now, if the clocks show the time according to the figure opposite?

Problem 6. Petya and Vasya played the naval battle on a 6 × 6 board. Vasya painted over the cells where Petya no longer has ships. Petya still has a three-deck ship ▢▢▢ . Which cell should Vasya attack to safely hit Petya's ship?

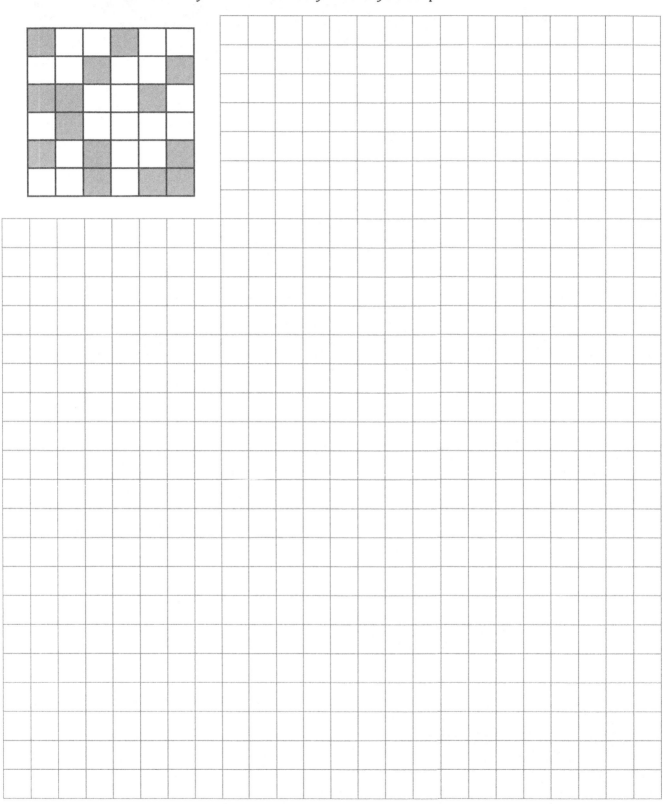

Problem 7. Tikhon represents the natural numbers of a single digit with the help of matches:

He also established the correct equality 2 + 6 = 9 − 1. Move 2 matches to get another correct equality.

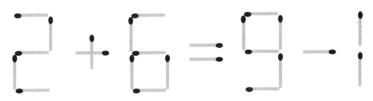

Problem 8. Vika, Nastya and Sonya are learning to count. Each of them has 1 or 2 candies. Sonya said, "We have at least 5 candies". Nastya: "Vika and I have equal amounts". Vika: "I have more candy than Sonya". It turned out that they were all wrong. How many candies does each have?

Olympiad 2018

(XXII Olympiad for Elementary School)

Problem 1. Color the four circles in three different colors so that two adjacent circles are different colors.

Problem 2. Indicate which of the 4 cube arrangements shown here are the same.

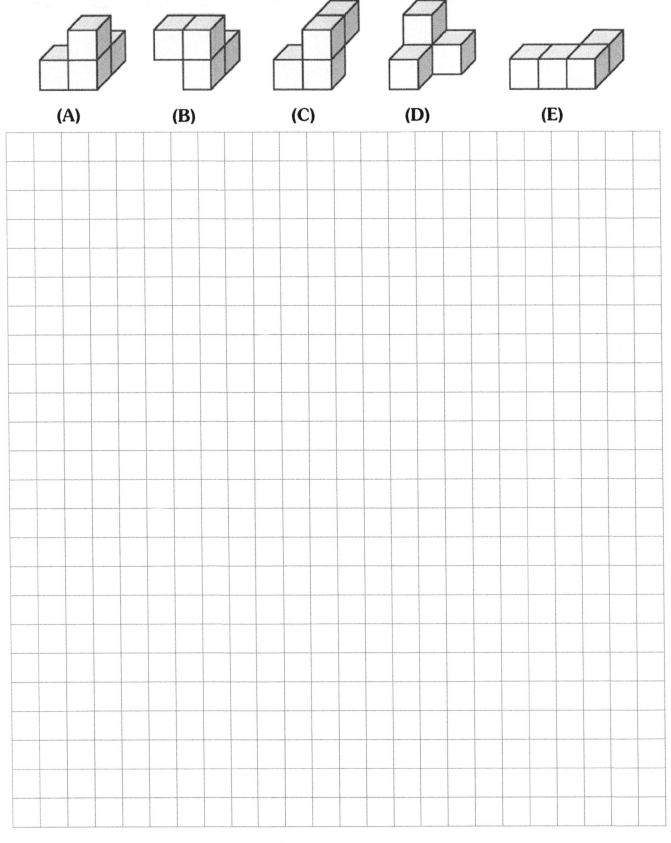

(A)   (B)   (C)   (D)   (E)

Problem 3. Varina's father is named Nikita Andreevich and her grandfather is Eduard Vasilyevich. What is the middle name of Varina's mother?

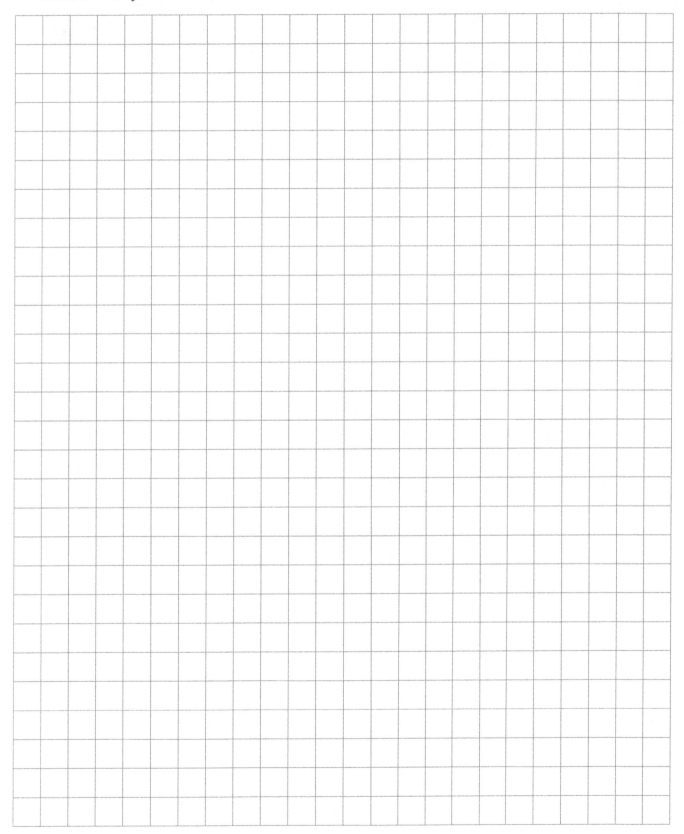

Problem 4. Jung is training to tie knots. The image shows his five attempts. What knots will be tied if the rope is pulled at the ends?

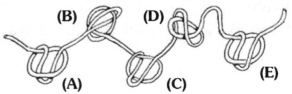

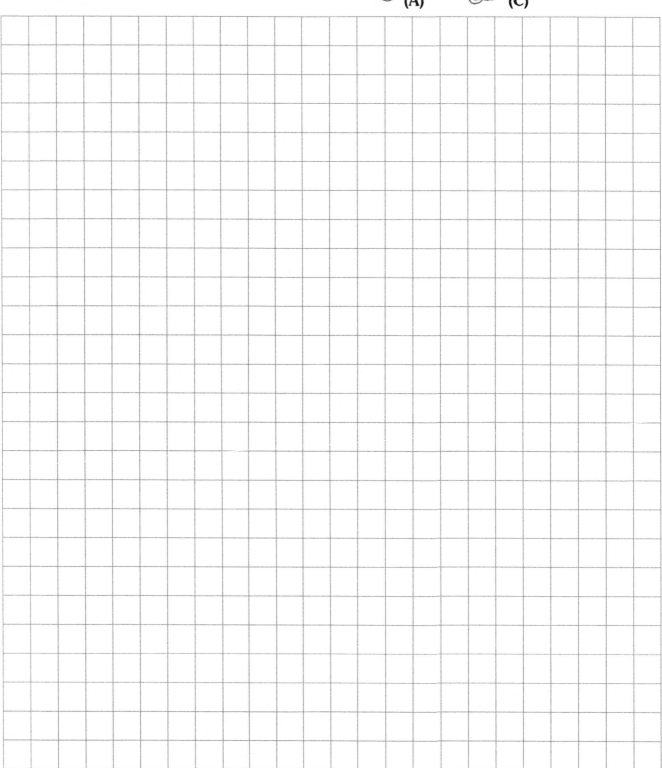

Problem 5. Masha forms numbers with matches. She places the number "5" in front of a mirror in *every possible way* and looks at it from different sides. What one- and two-digit numbers can she see this way?

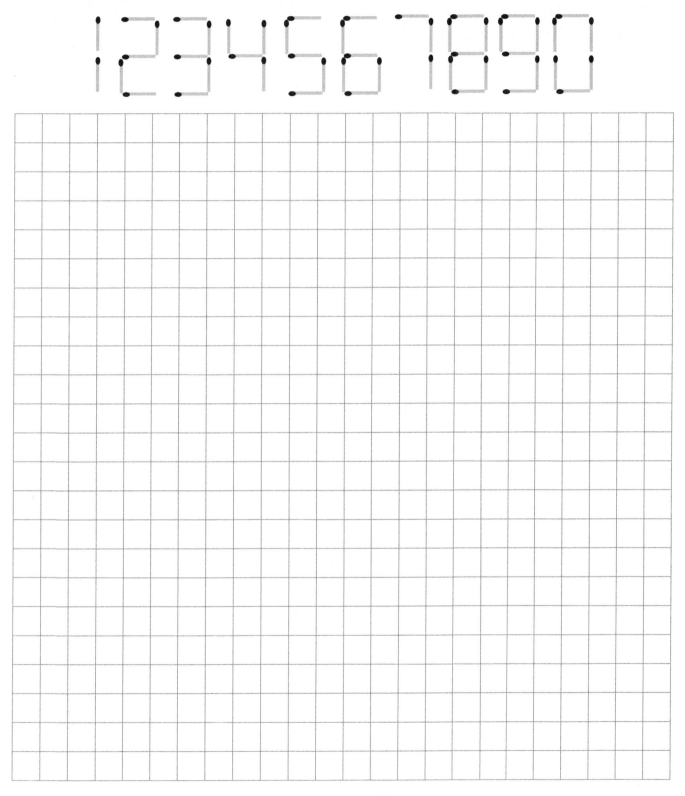

Problem 6. To play "Twister" a field of 12 cells with 4 different colors is used: red (R), yellow (Y), green (G) and blue (B) (see figure). Murka stretches across the field so that only 4 cells of different colors remain free, which are not adjacent. Indicate which cells Murka occupies.

R	G	B	Y
B	Y	R	G
R	G	B	Y

Problem 7. Piglet planted 10 acorns. All but three grew oak trees. All but two oaks have acorns. In all but one oak with acorns, the acorns have no flavor. How many oaks with tasteless acorns are there?

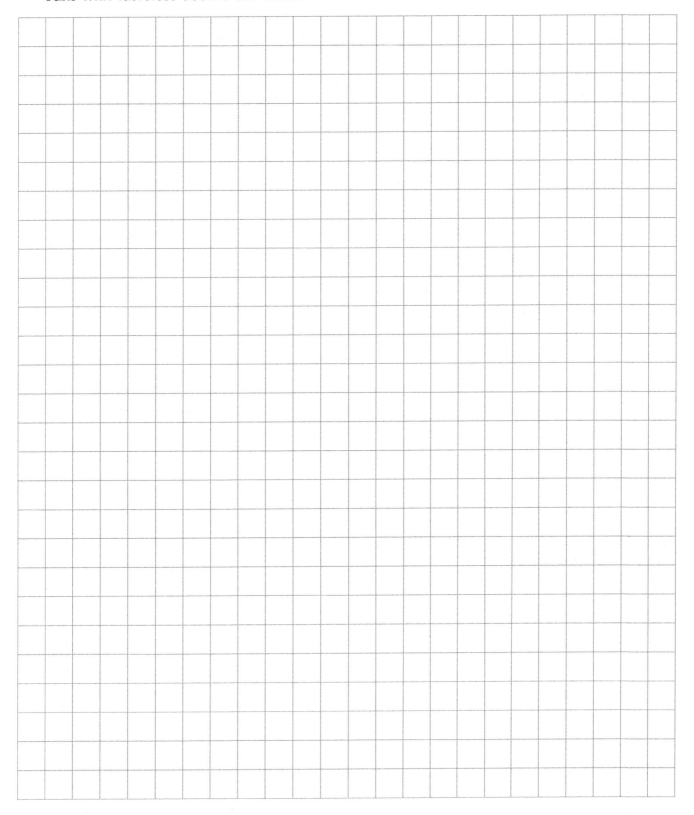

Problem 8. One day it was snowing all night. That night, three cars arrived at different times and parked near a house (as shown in the figure). Determine in what order the cars arrived.

Olympiad 2019

(XXIII Olympiad for Elementary School)

Problem 1. Replace the letters with digits (if different letters represent different digits) to get the correct equalities:

$$P - O = B - E = D - I = T + E = L + I$$

Problem 2. Three parrots have 9 nuts together. Red is 1 more than Green and Blue is 1 less than Green. How many nuts does each of the parrots have?

Problem 3. Winnie the Pooh flies with 7 balloons to look for honey and now he can't go down. Piglet has a pistol, from which he can only shoot 2 times. If he wants to pop several balloons in one shot. Show how Piglet can help Winnie the Pooh get him down by shooting all the balloons.

Problem 4. Alena wants to untangle her ribbons. How many are there?

1 2

There are 4 cirds

so I do 4 ÷ 2

4 ÷ 2 = 2

Problem 5. Buratino has four of the same coins. If he put them in 4 bags (see figure below) so that each bag contains a different number of coins. Indicate how the coins are distributed in the bags.

Problem 6. Natasha formed digits with matches and posted an incorrect equality as shown below.

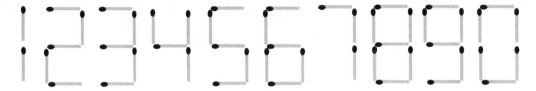

Move 2 matches to get the correct equality:

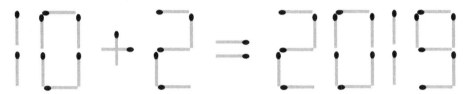

Problem 7. Go through the maze by turning left three times and right three times (in any order).

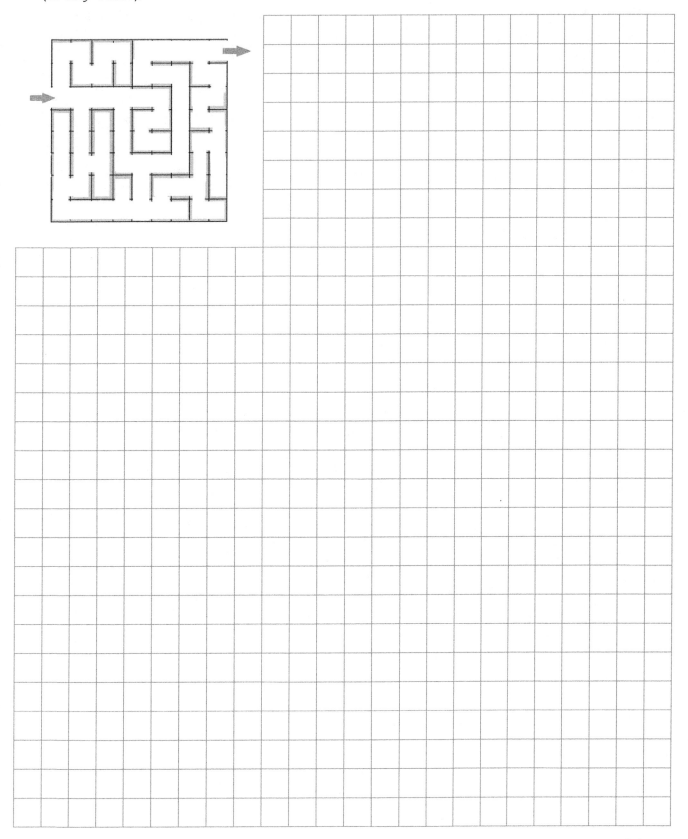

Problem 8. Five soldiers are standing in a row: Anton, Nikolay, Fedor, Alexey, Ivan. The colonel ordered two neighboring soldiers to break ranks, and then did the same with those whose names begin with the same letter. Finally, a soldier remained in line. What is his name?

Olympiad 2020

(XXIV Olympiad for Elementary School)

Problem 1. There are six brothers in the Zephyr family, two of them are twins, very similar to each other, like two drops of water. What are the names of these twins?

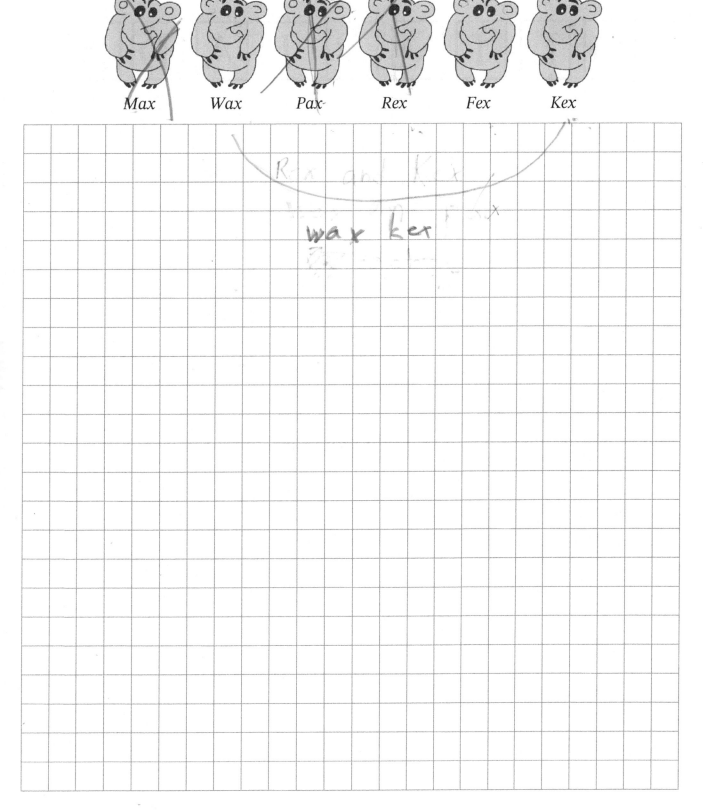

Max   Wax   Pax   Rex   Fex   Kex

wax kex

Problem 2. Masha formed a word with letters contained in gray and white cards. Grisha noted that it is possible to exchange white and gray cards so that the colors of the cards alternate. What cards must be exchanged?

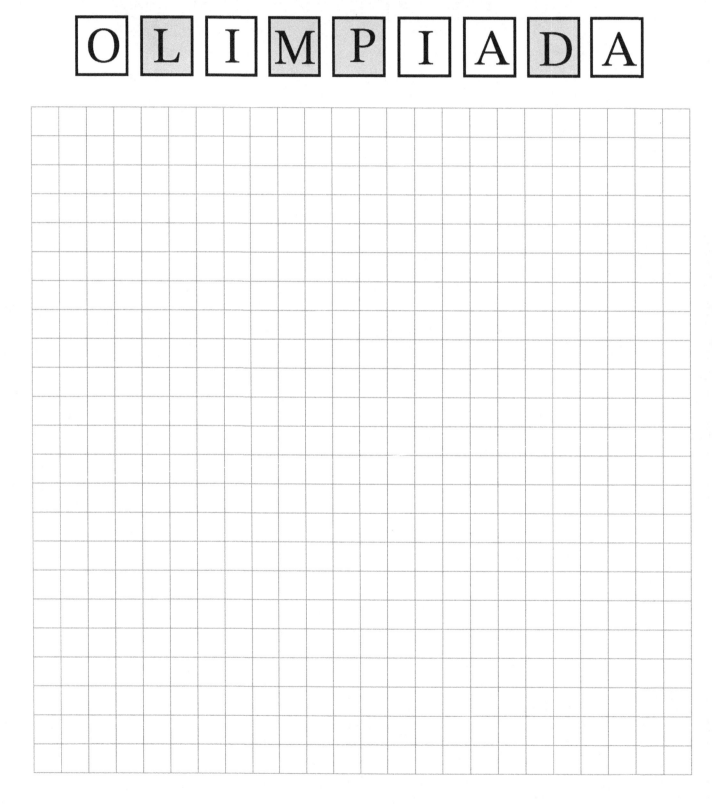

Problem 3. On Christmas Eve, Misha made Christmas trees from a honeycomb as shown in the figure on the left. From the figure on the right, she cut two Christmas trees. How did he do it?

Problem 4. There are several ♧ trees and ● light poles in the park. The light pole illuminates the trees strictly horizontally (↔) and vertically (↕). The tree closest to a light pole blocks the rest of the light in that direction. Mark which tree or trees A) are not illuminated with "O"; B) illuminated by exactly two light poles with "✕"

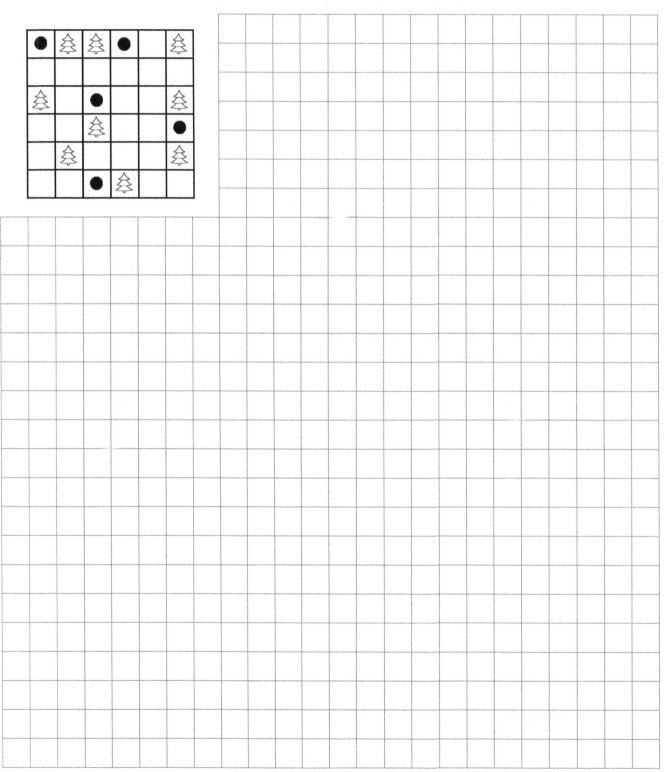

Problem 5. Misha took three dominoes 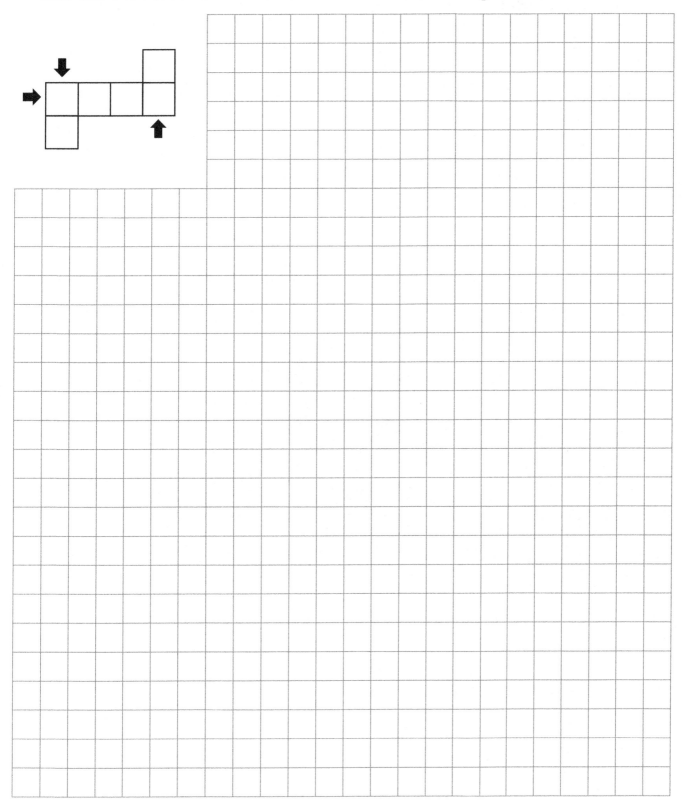 and put them in a scheme as shown in the figure. It turned out that the sum of points in two vertical lines and one horizontal is the same. Show how Misha arranged the dominoes.

Problem 6. The magic white sheet of paper changes its color to black in the places where the white parts touch. The square sheet was folded twice (as shown in the figure), pressed and cut diagonally. How many pieces that are completely black on both sides are there?

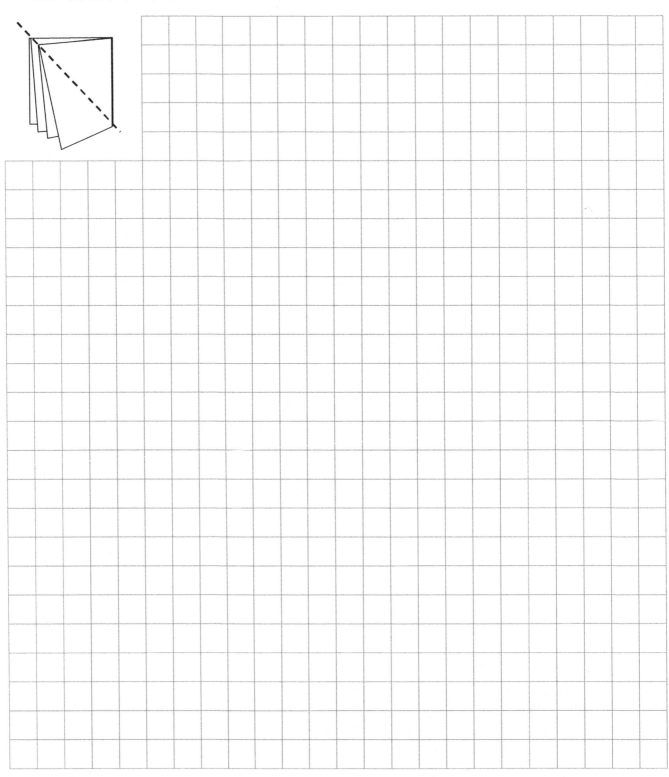

Problem 7. There are four houses along a straight street: blue, yellow, green, and red (in that order). The fox does not live in the red house. And the neighbors of the hare are a bear and a hedgehog. Who lives and where if the fox is not next to the hedgehog?

Problem 8. Snow was falling at night. In the morning, Fyodor in sneakers , Sharik in boots , and Matroskin in felt boots walked through the fresh snow. In what order were they?

Answers

Olympiad 2011

1. Nyusha.

2. Option (C).

3. 16 holes.

4. 2 green cells.

5. See the next figure:

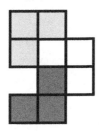

6. He should go into the thicket.

7. Winnie the Pooh.

8. Andrey loves oranges and Vera loves apples.

Olympiad 2012

1. Pavel Ivanovich.

2. The solution is the following:

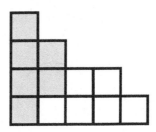

3. the picture of apple number 7 is missing. The figure shows the correspondences for the remaining fruits.

2   4   3   6   5   1

4. See the table below:

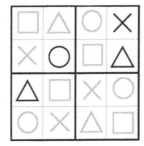

5. The Honda is blue, the Mercedes is yellow, and the Audi is red.

6. 8 kilometers.

7. Masha is 2 minutes faster.

8. Today is Saturday.

Olympiad 2013

1. A total of 18 legs.

2. Option (D).

3. The figure shows one of the possible cut options.

4. There will be four hands free.

5. Julia has a higher number in the mirror.

6. The Cheburashka route is: 3-4-1-3-5-2-1.

7. There is 12 russulas in the glade.

8. Hedgehog drove the car.

Olympiad 2014

1. There are 2 more pears than apples.

2. It rotates clockwise.

3. Tune in "AGA" at 12 o'clock and then tune in "OGO" at 15 or 17 o'clock.

4. Gosha will find 5 pieces after unfolding such a paper.

5. The teacher's name is Alisa Mikhailovna.

6. In the first car there are 2 kittens, in the second – 4 kittens, in the third – 1 kitten.

7. This number is 6243.

8. Mokhovaya Beard ate less than the others.

Olympiad 2015

1. 18 years old.

2. 796.

3. The second picture.

4. The solution is as follows:

5. By moving the two matches at the top and placing them at the bottom, we can then see 2 large and 1 small triangles.

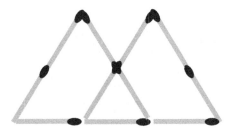

6. 7 ways.

7. 95 *cm.*

8. Vasya's yellow car arrived first.

Olympiad 2016

1. Option (D).

2. Anton is 8 years old, Borya is 9 years old, Kolya is 10 years old.

3. $21 - 9 - 9 = 3$ or $20 - 9 - 8 = 3$ or $20 - 8 - 9 = 3$.

4. In the figure, the possible cells are shaded gray.

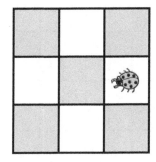

5. 8 pieces.

6. Kasimir drew the black square, Vasya drew the gray circle, Gosha drew the white triangle.

7. The correct equality is $40+18=58$.

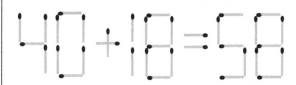

8. Fire truck - Yegor, dump truck - Tikhon, excavator - Vitalik.

Answers

Olympiad 2017	Olympiad 2018

Olympiad 2017

1. Option (D).

2. Petya is faster.

3. For example, $1 < 6 < 7 > 2 < 3 < 6 > 5 > 4$.

4. Masha, Yegor, Katya, Tikhon, Sveta.

5. It's 6 o'clock now.

6. The required cell is marked in the figure.

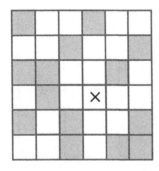

7. The correct equality is 2+5=6+1.

8. Sonya has 1 candy, Nastya has 2 candies, and Vika has 1 candy.

Olympiad 2018

1. For example, red - blue - yellow - red.

2. The options (A) and (D) are identical.

3. The middle name is Eduardovna.
Since Nikita Andreevich's middle name is not Eduardovich, Eduard Vasilyevich is not his father. But he is Varina's grandfather. This means that this grandfather is the father of Varina's mother, from which we get her middle name.

4. Knots (C) and (E) will be tied.

5. Masha will be able to see the numbers 5, 2, 3, 25, 52.
The dotted line shows the position of the mirror:

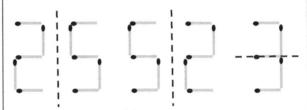

6. Two symmetrical variants are possible:

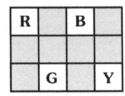

7. There are 4 oaks with tasteless acorns.

8. Car B arrived first, Car A arrived second, Car C arrived third.

Olympiad 2019

1. For example, $9 - 3 = 8 - 2 = 7 - 1 = 4 + 2 = 5 + 1$.

2. Green has 3 nuts, Blue has 2 nuts, Red has 4 nuts.

3. Two shots in the shaded areas will suffice, one in each.

4. 3 ribbons. In the figure, each ribbon is highlighted in its own color.

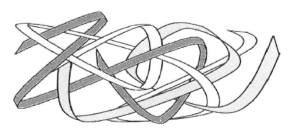

5. An example is shown in the next figure:

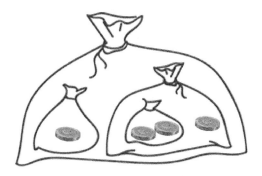

6. An option is shown in the next figure:

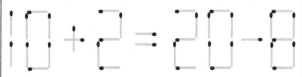

7. An option is shown in the next figure:

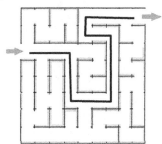

8. The name of the remaining soldier is Ivan.

Olympiada 2020

1. Wax and Kex.

2. "P" and "I" must be interchanged

3. An example is as shown as follows:

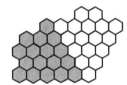

4. The trees are marked on the image.

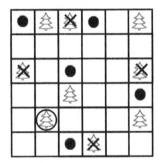

5. Un example is shown in the figure:

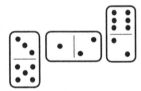

6. 2 pieces.

7. The fox is in the blue house, the bear is in the yellow one, the hare is in the green one, and the hedgehog is in the red one.

8. First Matroskin, then Fyodor, and last Sharik.

Made in the USA
Las Vegas, NV
13 May 2024

89884118R00070